Understanding the Elements of the Periodic Table™

COPPER

Paula Johanson

rosen
central™

The Rosen Publishing Group, Inc., New York

*For those who gave me a sense of wonder in science
and the natural world*

Published in 2007 by The Rosen Publishing Group, Inc.
29 East 21st Street, New York, NY 10010

First Edition

Library of Congress Cataloging-in-Publication Data

Johanson, Paula.
Copper / Paula Johanson.
    p. cm.—(Understanding the elements of the periodic table)
Includes bibliographical references and index.
ISBN 1-4042-0706-6 (library binding)
1. Copper—Popular works 2. Periodic law—Popular works.
I. Title. II. Series.
QD181.C9J64 2007
546'.652—dc22

                                                            2005028997

Manufactured in the United States of America

CPSIA Compliance Information: Batch #210230YA: For Further Information Contact Rosen Publishing, New York, New York at 1-800-237-9932

On the cover: Copper's square on the periodic table of elements. Inset:
The atomic structure of copper.

Contents

Introduction

Has anyone ever offered you a penny for your thoughts? This small coin is made mostly of copper, a metal that people have used since our early ancestors found shiny pieces of it among riverbed stones and sometimes in the ashes of their fires. When humans began chipping rocks into tools and arrowheads, they also used pieces of copper for tools. Copper was softer than many rocks but still useful.

Metallurgy is the study of metals and their chemistry. It is also the science that deals with extracting metals from their ores, of purifying or alloying (mixing) metals, and of creating useful objects from metals. Metallurgy probably began with the early use of copper. Copper melts at a lower temperature than iron, so copper was probably the first metal our ancestors learned to smelt out of ore in simple kilns and bonfires. Copper ores are often colored, making them easier to recognize than ores of most other metals. People eventually learned that molten copper also combines with other metals such as zinc, tin, or nickel to make harder alloys, which can be used for making weapons and farming and construction tools. Copper is also present in plants and makes up a tiny but important part of our food and our bodies.

Copper, silver, and gold are called the coinage metals because they are used for making coins. Coins often had an image of a god or ruler stamped onto one side. Modern coins from Canada and Great Britain

Copper has always been found in nuggets in streambeds *(above)*. Copper can also be mined from rock in Earth's crust, where it is combined with other elements in copper ores. The heat of a simple kiln or a bonfire is enough to melt copper out of ore, unlike some other metals that melt at a higher temperature.

show the face of Queen Elizabeth II with these words in Latin: *D.G. Regina*, or "Queen by Grace of God." The ancient Greeks associated copper with Aphrodite, goddess of love and wife of Hephaestus, the Greek god of fire and metalworking.

Copper coins are still associated with reverent thoughts. People throw pennies into fountains while making a wish. Some native North Americans when traveling by ship even throw a penny into the water as their journey begins.

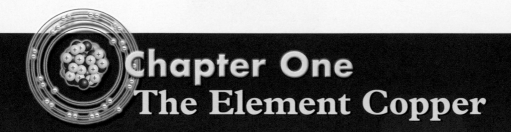
The element copper is a familiar part of life in the developed world. Copper's chemical symbol, Cu, comes from the Latin word *cyprium*, which was later shortened to *cuprum*, meaning "from the island of Cyprus." Copper found in its pure state in nature, called native copper, has been known since ancient times. Tools made of copper alloys (mixtures of copper with other elements) have been used since the beginning of the Bronze Age, around 4000 BC. In its pure form, copper is used in electrical wire, plumbing pipe, and jewelry. Pure copper is a shiny metal with a red-orange color. The surface of pure copper tarnishes very slowly when exposed to air and water, but this thin green or brown patina does not crumble like rusty iron. (A patina is a film that forms naturally on copper or its alloys as a result of oxidation.) Copper is a relatively soft metal that can be bent, beaten, or cast into useful shapes. It is even more useful when it is combined with other elements, usually tin or zinc, to make bronze or brass alloys. Bronze and brass are used to make many tools, screws, and other hardware that resist corrosion, or decay.

Atoms, Elements, and Compounds

Everything you can see in the world—and nearly everything in the universe—is made up of elements. Some objects are pure, with one kind of element only. Some are compounds, with different elements bonded chemically, sticking together as if they were glued together in different arrangements and with glues of different strengths.

How small a piece can you have of an object? If you started with a trumpet, you could see that it has some brass metal parts and some steel parts. The smallest piece of brass you could cut would have copper and zinc, which were originally melted together at a high temperature. You could use chemical reactions to separate out the copper. The smallest portion that you can have of an element is called an atom, which is far too small to see with the microscope you might use in school. Atoms of copper are so tiny that about thirty-five million of them, side by side, would make a line only 0.4 inch (1 centimeter) long. One atom would be the smallest piece of copper you could ever have. Inside that atom are even smaller pieces called subatomic particles.

Inside the Atom

Atoms are made of three main kinds of subatomic particles. At the center of the atom, the nucleus is a dense core of protons, which carry a positive electrical charge, and neutrons, which have no charge. The nucleus is a small part of the atom—it's like a marble in the center of a football stadium. But the nucleus contains nearly all the mass of the atom. Copper has twenty-nine protons in its nucleus. That is how the element is identified. Any atom, anywhere in the universe, with twenty-nine protons in its nucleus, is an atom of the element copper.

Around the atom's nucleus, the electrons occupy overlapping layers called shells. Electrons have a negative electrical charge and are attracted to the protons in the nucleus. The negative and positive electrical charges of the atom are balanced, and the number of electrons and protons are equal. An atom of copper has twenty-nine protons and twenty-nine electrons.

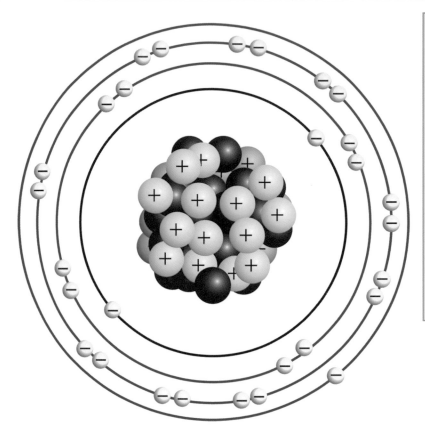

All atoms are made up of the subatomic particles protons, neutrons, and electrons. The protons and neutrons cling together in the nucleus. The electrons circle the nucleus in layers called shells. It's easier to look at a drawing of an atom like this one with the electrons shown close to the nucleus, but a more realistic model would have tiny electrons whizzing around a stadium with a raspberry at the center for the nucleus.

In metals, the outermost electrons are less tightly held to the nucleus than the inner electrons. These outer electrons are called valence electrons. They can be readily exchanged with neighboring atoms, so that electrons move from one atom to the next. This flow of electrons is what enables metals to conduct heat and electricity. When an atom of metal gives up one or more of its valence electrons to an atom of nonmetal, both of these atoms become charged particles called ions. The metal ion is called a cation. It has a positive charge because it has lost one or more electrons.

The Periodic Table

Some elements, such as copper, gold, iron, and mercury, have been known since ancient times. As elements have been discovered and studied over the years, scientists have organized this knowledge. Elements are

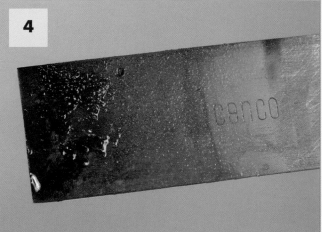

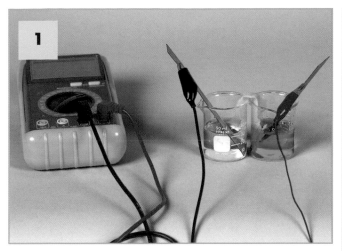

A solution of copper sulfate ($CuSO_4$) and water can be used to investigate how elements can be released from compounds and electroplated. When water is added to the copper sulfate, the copper sulfate is broken into copper ions and sulfate ions. The copper ions look for a negatively charged ion with which to combine. By using a battery source and a copper (Cu) strip that is connected to the anode and a zinc (Zn) strip that is connected to the cathode, and the other ends of both strips placed into the solution, the electroplating experiment can be speeded up. Eventually, some of the copper from the solution will plate onto the copper strip.

Copper 29 Cu 64 Snapshot

Chemical Symbol:	Cu
Classification:	Transition metal
Properties:	Reddish-orange color; solid at room temperature; malleable; ductile; conductor of heat and electricity
Discovered By:	Was known by ancient civilizations
Atomic Number:	29
Atomic Weight:	63.546 atomic mass units (amu)
Protons:	29
Electrons:	29
Neutrons:	36 or 34
Density at 68°F (20°C):	8.96 grams per cubic centimeter (g/cm³)
Melting Point:	1,984°F; 1,084°C
Boiling Point:	4,644°F; 2,562°C
Commonly Found:	Earth's crust as copper ores; also as native metal

arranged on a chart called the periodic table of elements. Russian chemist Dmitry Mendeleyev (also spelled Mendeleev) was the sixth person to publish a chart of the elements, later called a periodic table. Mendeleyev made his chart in 1869 and used it to teach his students at the University of St. Petersburg in Russia. Mendeleyev's version of the chart was the one that became most widely known.

The elements are organized in the periodic table by increasing atomic number (the number of protons in the nucleus of an atom of an element), and by trends or patterns in reactivity among the elements. (Reactivity refers to how easily the elements interact with other substances.) The most obvious distinction in the periodic table is the difference between the metals (in the left and bottom parts of the chart) and the nonmetals (in the right and upper parts of the chart). A staircaselike line separates the elements in the table. Metals are located to the left of this line.

Most metals have similar qualities, such as the ability to conduct electricity. They can be polished to be shiny. Most metals are malleable, which means they can be pounded into shapes without breaking. Malleable metals are usually also ductile, which means they can be pulled into wires. Metal atoms are also able to give up one or more electrons from their outer electron shells to form ions with a positive electrical charge.

The nonmetals are grouped together to the right of the staircaselike line. Among these nonmetals are the noble gases, which do not react with any elements under most conditions, on the far right of the chart. The other nonmetals are able to pick up one or more extra electrons; this gain of electrons forms ions with a negative electrical charge (called anions). Metalloids such as boron and silicon behave like nonmetals in most respects, but they conduct electricity, though less well than metals. They are located on the immediate right of that staircaselike line, between metals and nonmetals.

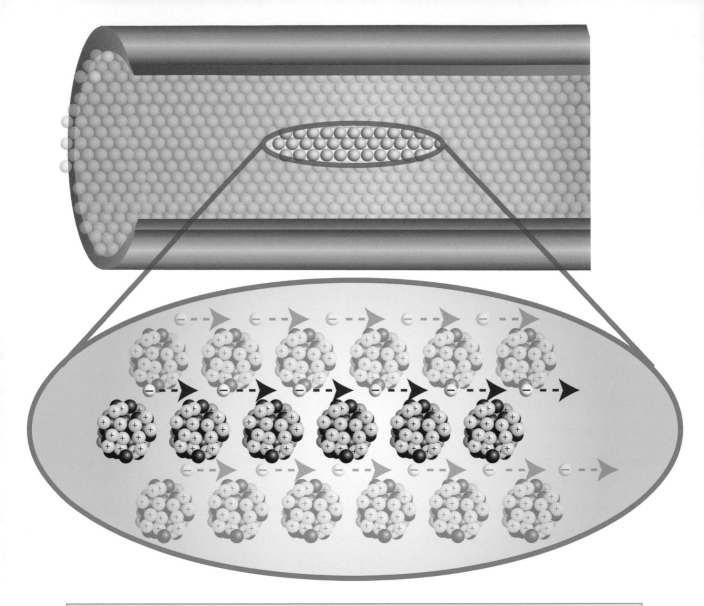

Inside a piece of copper wire are atoms, too small to see. Each nucleus is surrounded by its own shells of electrons. The outermost shell of valence electrons can move on to a neighboring atom, pushing another electron on to the next atom. This easy movement conducts heat through all parts of a metal object. If extra electrons are made available to one end of a piece of copper wire, the electrons flow through the wire, moving from atom to atom.

Copper is located on the left of the line, among the transition metals. Transition metals are somewhat less reactive than the alkali metals (lithium, sodium, potassium, rubidium, cesium, and francium) on the far left of the chart. They also have similar characteristics: they are malleable, ductile, and conduct heat.

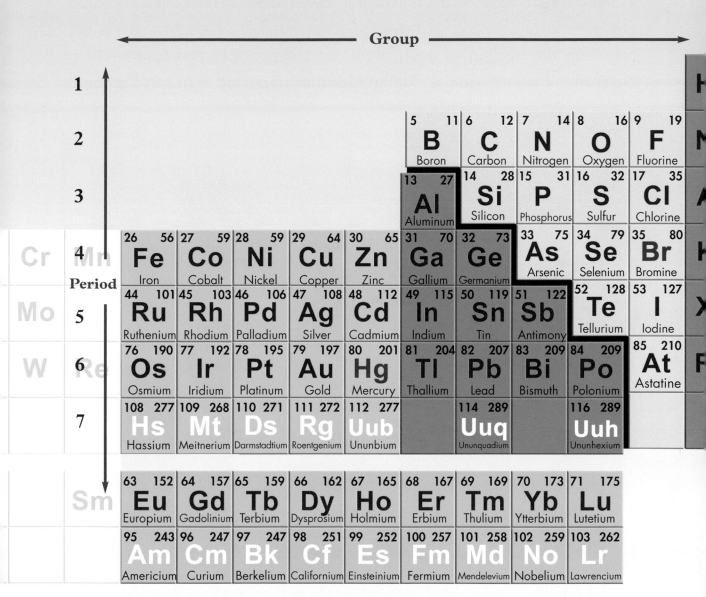

The periodic table sorts elements into groups and periods. Each element in a group has similar properties. Each element in a period has one more proton than the element on its left, and one fewer than the element on its right; this affects its physical and chemical properties. Studying these properties helps us understand the relationships between the elements. The staircaselike line divides metals from nonmetals. Metalloids, such as boron (B) and silicon (Si), are on the immediate right of the staircase line.

Periods and Groups

In the periodic table, each horizontal row of elements is called a period. Elements in one period all have the same number of electron shells surrounding the nucleus of each atom. Because copper has four shells of electrons, it is in period 4. All the elements in period 4 are in the fourth

How much difference can one proton make? Plenty! Each of these samples of metal (clockwise from left: aluminum pellets, nickel-chrome ore, nickel bars, titanium bars, iron-nickel ore, niobium bars, and chromium granules) is slightly different from the sample of copper in the middle. Some are shinier or different in color or hardness. The number of protons defines the element and determines its properties. Even one proton more or less can make one element very different from another.

row down from the top of the chart. As you read across the row, each element has one more proton in its nucleus than the element on its left, and one more electron in its outer shell.

The vertical columns of elements are called groups. Elements in a group have similar properties. Copper is in group 1B (also called group 11) with coinage metals silver and gold. All the elements in group 1B are in the eleventh column from the left side of the chart. Copper has four electron shells, silver has five, and gold has six. Much like group 1A elements, group 1B elements are able to give up one electron easily, but they oxidize more slowly than group 1A elements and are more likely to form complex molecules.

Unique Qualities of Elements

An element is defined by the number of protons found in the nuclei of its atoms. Any atom with twenty-nine protons in its nucleus is an atom of copper. In the periodic table, you can see the atomic number 29 written above and to the left of the symbol for copper (Cu). An atom with thirty protons would be a different element (zinc), with different properties. An atom with twenty-eight protons would be another element, nickel. The presence of one proton more or fewer makes a big difference: nickel is shiny and silvery, copper is shiny and reddish orange, and zinc is duller and gray. Unless the atom is an ion, the atomic number also indicates the number of electrons.

Atomic Weight

In the periodic table, you can see the number 64 written above and to the right of the symbol Cu for copper. That is the atomic weight of copper—the number of protons plus the average number of neutrons found in an atom of that element. Atomic weight is also known as atomic mass, and it is measured in atomic mass units (amu). The atomic weight

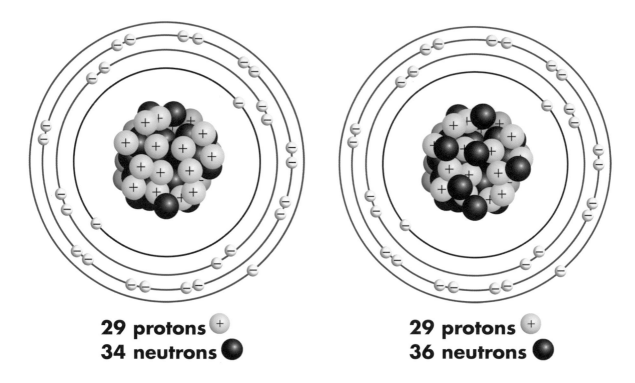

29 protons +
34 neutrons ●

29 protons +
36 neutrons ●

Atoms of copper are fairly complex. Each contains twenty-nine protons in its nucleus. Twenty-nine electrons surround the nucleus. There are either thirty-six or thirty-four neutrons in the nucleus of an atom of copper. Both of these isotopes of copper are stable and are commonly found in any sample of copper ore or smelted metal.

of copper is 63.546 amu, which has been rounded to two digits, 64, in our periodic table.

Atomic weight is an average weight and not an exact weight because elements are not always made out of identical atoms. Some copper atoms (69 percent) have 34 neutrons, while others (31 percent) have 36 neutrons in the nucleus. The copper atom with 36 neutrons adds more weight or mass to that atom's nucleus than does the one with 34 neutrons, but it does not affect the element's chemical properties. Many elements have two or more possible numbers of neutrons in their nuclei, and these alternate versions are called isotopes.

Chapter Two
The Properties of Copper

Each element can be described by its chemical and physical properties. These characteristics are what scientists use to identify and classify an element. We can observe the physical properties of an element by itself, without mixing it with anything else. The physical properties that can be observed include an element's hardness, density, and conductivity. We can also observe the element's phase at different temperatures. Is it a solid at room temperature? At what temperature does it melt into a liquid? What is the boiling point?

On the other hand, the chemical properties of an element describe its abilities to form new molecules with other elements. One kind of chemical property is copper's ability to combine with oxygen to form oxides. This can only happen at very high temperatures, when the copper is red-hot. Some other metals oxidize very easily. Observing the chemical properties of an element means observing as the element is changed from a pure substance of one kind of atom to a new substance, a compound of two or more kinds of atoms that are chemically bonded into a molecule, fitting together as if the atoms were glued together in various ways. For example, when copper gradually reacts with oxygen and sulfur in the air, it changes from a soft, reddish-orange, shiny metal to form a brittle, greenish crust on the surface.

Copper's Phase at Room Temperature

An element is found in one of three phases at room temperature: gas, liquid, or solid. Knowing the physical state of an element at room temperature helps scientists to identify it. Copper is a shiny, red-orange solid at room temperature. A solid keeps its shape and volume, and resists changing that shape under pressure or compression.

Copper can be shaped into pipes or wires and other useful things, with confidence that these tools will not quickly rust away or corrode. Because copper is malleable, it can be bent and shaped without cracking. Copper is also ductile and can be made into a round, thin wire that is thinner than a human hair.

Copper's Density

Density is a measure of how much mass an object contains in a given volume (mass per unit volume, frequently expressed as grams per cubic centimeter). Copper's density is 8.96 g/cm^3. Solids are usually denser than liquids, and liquids are denser than gases. If you drop a solid piece of copper into water, it will sink because copper is denser than water.

Copper's Hardness

Pure copper is a relatively soft metal. This is a useful quality, allowing copper to be shaped. To make it harder, other elements can be added, but if large quantities of the other elements are added, the resulting alloy is a little more vulnerable to corrosion.

The hardness of common substances is measured with Mohs' scale. Friedrich Mohs was a German mineralogist. He learned from miners who routinely used scratch tests to compare minerals. In 1822, he published a scale comparing ten groups of substances, ranked in order of increasing hardness. Substances from 1 to 2.5 on his scale are considered soft. Those between 2.5 and 5.5 are of an intermediate hardness, and substances at 5.5 and above are considered hard. The hardest substance of all, diamond, is in the tenth group. On Mohs' scale, copper has a hardness of 3. Your fingernail has a hardness of 2.5, and a piece of quartz rock has a hardness of 7. A copper tool could scratch your fingernail, but a quartz

Mohs' Scale
Hardness Rating Examples

1	Talc
2	Gypsum (rock salt, fingernail)
3	**Calcite (copper [Cu])**
4	Fluorite (and iron [Fe])
5	Apatite (and cobalt [Co])
6	Orthoclase (and rhodium [Rh], silicon [Si], tungsten [W])
7	Quartz
8	Topaz (and chromium [Cr], hardened steel)
9	Corundum (sapphire)
10	Diamond

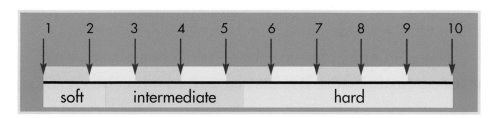

From ancient tools, we learn about their makers and users. This bronze pot from ancient China, made of copper and tin and probably cast in three sections, has three legs, for standing on uneven ground instead of a smooth floor. A sturdy stick or metal rod would fit through both handles if it had to be moved when hot. The decorations show this useful pot was also meant to be admired.

rock could scratch the copper. A copper knife would be hard and sharp enough to cut meat and vegetables, though it might be too soft for many other uses. Other elements can be added to copper to make harder alloys.

Bronze is an alloy of copper that can contain up to 25 percent tin. It can be cast or beaten into tools that are harder than copper and better at keeping a sharp edge. This discovery was made 5,000 or 6,000 years ago. The era when bronze tools were commonly in use around the Mediterranean is known as the Bronze Age.

In the Greek epic poems *The Iliad* and *The Odyssey*, composed more than 3,000 years ago by the poet Homer, bronze swords and bronze-tipped spears were the weapons of war. Homer, who was blind, usually described characters injured in the battles as being "gashed with the mangling bronze."

Brass is an alloy of copper that can contain from 5 percent to 45 percent zinc. It is easily molded or beaten into useful shapes and is much harder than pure copper. Some brass alloys look like gold, but it is easy to tell brass from gold. A brass bracelet, for example, would weigh much less than an otherwise identical gold bracelet.

Brass is used in musical instruments, partly because it contributes to a pleasant tone in instruments, it is attractive, and it isn't heavy. Brass is also

Brass has given its name to an entire section of an orchestra. These musical instruments, which include the French horn, saxophone, and trumpet, have a brighter tone than did horns in the Middle Ages that were carved out of animal horn and wood. Brass and bronze have been used to make horns and cymbals since ancient times. Although its color is similar to that of gold, brass is an alloy of copper and is also made of zinc.

used in musical instruments because it is not as vulnerable to damage by temperature and humidity changes as some other metals or alloys.

Brass is made into screws and other hardware because it resists corrosion. Look around any building and you will probably find brass door hinges, doorknobs, or handrails. Travelers' trunks are fitted with brass handles, corners, and locks. Most of the keys used to open locks or operate motor vehicles are made of brass.

Many of the metal fittings on boats and ships are made of brass to resist corrosion from freshwater or salt water, but boat makers never fasten sheets of iron alloys with brass screws or bolts! If two different alloys touch where water will be present, this sets up a weak electric current. The copper alloy becomes the cathode and the iron alloy becomes the anode just like in a battery. The water has enough ions and oxygen

Bronze in Statues

Some artists, such as Bill Reid, rub black wax polish over their bronze statues to discourage even the slightest weathering. Reid's statue *The Jade Canoe* is located at Vancouver International Airport in Canada, where portions of the dark, greenish canoe and its mythical passengers have been rubbed shiny by the hands of many travelers. An image of *The Jade Canoe* was put on the Canadian twenty-dollar bill in 2004.

dissolved in it to allow this effect to happen. The result will be a corroded hole in the iron alloy sheet and a brass bolt with a brittle plating of minerals coating it. In salt water, this can happen in a matter of days, as amateur boat builders may be surprised to learn.

Copper can be mixed with both tin and zinc to form bronze or brass. You may have seen statues cast in these alloys, which can be polished to keep a bright golden color or allowed to take on a dark brown or greenish color when the outer surface combines with compounds in the air to make a layer of carbonates and sulfates.

Copper: Conductor of Electricity and Heat

Copper conducts electricity and heat, as all metals do. Its valence electrons, electrons in the outer shells of copper atoms, are able to move about from one copper atom to another. This movement of electrons is what we call electricity. This movable sea of electrons in metals also makes them good conductors of heat, as the heated electrons move about more quickly and distribute the heat throughout the piece of metal.

Among commonly used metals, copper is the best conductor of heat and electricity. Many chefs prefer to use fine cooking pots made entirely

Copper is an excellent metal to use in cooking pots because of its good conductivity, but the metal will discolor when touched by an acid food like tomatoes or vinegar. Copper pots are usually lined with another metal, such as tin, steel, or aluminum, which does not discolor. Copper pots also provide even heating for the cooking of foods. However, copper pots can be heavy and expensive to purchase, and they might require some retinning over time.

or partly of copper. Cast-iron frying pans are more common only because iron is cheaper than copper. Chef Julia Child once tested a solid gold frying pan for a scientist and found it even better than her copper pans because it seasoned well with grease and held a great deal of heat and released it gradually. However, a $10,000 pan big enough to cook just one egg is not as practical as a copper pan.

Even though iron is cheaper, copper is used in electrical wire because it conducts electricity better and it doesn't rust. Silver is the only metal that conducts electricity better than copper, but it's not as common, and the price of silver would make electronic devices too expensive for most households and businesses.

Like gold and silver, copper is used in making medicines and machinery as well as coins and jewelry. If you have a gold coin or a piece of gold jewelry that is not twenty-four karats (100 percent gold), there is some copper or silver blended with the gold.

Copper's melting point is 1,984 degrees Fahrenheit (1,084 degrees Celsius). Some wood fires can burn that hot. Thousands of years ago, people discovered lumps of copper metal in the ashes of their bonfires, melted out of the rocky ore under their fires. Because copper melts at a lower temperature than many other metals, it can be used to make fuses, which melt when too much heat or electricity passes through them. A fuse is put into an electric circuit to protect the rest of the device from damage. Technicians prefer to open a panel and change a fuse as easily as changing a battery in a radio, instead of hunting through all the wires and hard-to-reach connections looking for a spot ruined by too much heat or electricity. A ship's captain would rather replace a cheap little fuse than discover that the entire radar unit is ruined.

Copper is found throughout the world, in the rocks of the earth's crust. There is copper in meteorites that fall to the earth from space, too. From studying the light of stars and the reflected light from planets and asteroids, scientists know there is copper elsewhere in the solar system and the universe. For every billion atoms in the universe, one is copper. That doesn't seem like much, but most of the atoms in the universe are hydrogen and helium in big clouds of gas and in stars. Almost all the copper in the universe would be found in the rock of planets and asteroids.

Copper in the Old World

As stated earlier, the word "copper" is derived from the Latin *cuprum*, which means "from the island of Cyprus." This island has extensive copper deposits, and for thousands of years it was a primary source of copper in the Mediterranean area. In ancient Greek mythology, Cyprus was the home of the goddess of love, Aphrodite, and copper was the metal associated with Aphrodite.

The beginnings of modern science and metallurgy are linked to old myths through the alchemists. These people began studying the elements and minerals of the world with two purposes in mind. The first was to

Space Probe

Deep Impact, the spacecraft that the National Aeronautics and Space Administration (NASA) sent to comet Tempel 1 on July 4, 2005, released a probe with a camera to collide with the comet. Half the mass of the probe was 660 pounds (300 kilograms) of copper. There isn't much copper in comets naturally, and copper reacts slowly with other elements so it would not interfere with scientists being able to tell from photographs what elements are in the comet.

This is an artist's view of the moment that *Deep Impact*'s probe collided with comet Tempel 1. The *Deep Impact* spacecraft had cameras to observe the probe's collision. From the color of the explosion, scientists are able to tell what elements are in the comet—a dirty snowball black as coal.

understand the nature of the world and, through it, spiritual ideals. The second purpose was to acquire those spiritual ideals and also their worldly counterpart: the philosopher's stone, which would turn lesser metals, such as lead and zinc, into gold, and make people younger. Copper was considered almost as fine as gold because it does not rust away like iron. In their writings, the alchemists used the symbol of Aphrodite and the planet Venus as a symbol for copper. This symbol is still used today as a symbol for female. The word for alchemy came from Arabic and eventually found its way into modern English as the word "chemistry."

The modern symbol for female has also been used as a symbol for the goddess of love, the planet Venus, and the element copper. It may be one of the oldest written symbols, and it is still in use today.

Copper Ores

Copper can be found as deposits of ore in sulfides, arsenides, chlorides, and carbonates. These ores are often colored, so the deposits can be easy to recognize. There are useful deposits of copper in Canada, Chile, the Democratic Republic of the Congo, Germany, Italy, Peru, the United States of America, and Zambia. You may have seen a rocky cliff stained by rainwater running over copper ores. There are usually copper ore samples, with formal names such as chalcopyrite, or informal names such as peacock ore and fool's gold, on display in hobby stores and museums. Some ores, such as malachite, are beautiful semiprecious stones.

Copper is also found as pure metal or native metal, sometimes known as native copper. These useful nuggets found in streambeds were probably the first pieces of metal to be recognized and used by people thousands of years ago. In most parts of the world, copper nuggets are more common than gold.

Copper in the New World

In North America, there were extensive deposits of both copper ores and native copper, which were used by Native Americans and traded

This open mining pit shows how the copper ore has been removed. The ore is carried in truckloads up the ramps and taken to the refinery. Rainwater takes up minerals from the ore and collects at the bottom of the pit. It may need to be treated or dumped into deep shafts in order not to poison surface water. After the mine is exhausted, even if the cover of rock and soil is replaced, it can take decades to grow a new natural forest over the mine.

throughout the Americas. In Arizona, copper-bearing minerals were used to make body paint and to decorate pottery, as well as to make tools and weapons. The capital city of Canada's Northwest Territories was named Yellowknife by European fur traders because of the copper knives used by the Déné people. Currently, there are few deposits of native copper remaining in the world; most of the copper mined today is taken from the ground as copper ores.

Copper was also used by Native Americans on the west coast of North America (in present-day British Columbia, Canada) during the

This copper's unusually simple design of a grizzly bear may hide a long history of use and re-use. Like silver and gold, copper ornaments have often been enjoyed by a series of owners and remade again and again. This ornament in the Royal British Columbia Museum in Victoria may have been made from copper gathered many hundreds of years ago or only decades ago.

nineteenth and twentieth centuries to create a shield-shaped ornament called a copper, or *tlakwa*. These coppers were too soft to be effective as real shields in war. Instead, a copper would be decorated with images from a clan's lineage, similar to the designs on wooden boxes, totem poles, and woven blankets. Coppers were traded for blankets and food, which could then be distributed to many guests at a grand gift-giving called a potlatch. Wealthy hosts gave away a copper and much or most

Copper ores can be beautiful semiprecious stones as well as sources of metal for sculpture. Malachite, which is pictured here with green crystals of the mineral dioptase, can be polished to make fine jewelry and ornaments worth more than the copper that could be extracted from the ore. Some pieces need little work to show their natural beauty.

of their immediate family's accumulated wealth to relatives, friends, neighbors, and rivals. Coppers might be traded or sold several times at one gathering or over a number of years. Potlatches were an effective way of distributing practical wealth (such as food and blankets) and of exchanging practical wealth for a symbol such as a copper or the inherited right to perform a song or dance.

The copper mining industry is thriving today, particularly in Arizona. This mineral was essential at the beginning of the industrial age, and today it is necessary in the production of computers. For just one example: if you went to a hydroelectric dam to look at the dynamos that are turned by water, you would see that these dynamos have great coils of copper wire turning around a magnet at the core. That generates a flow of electricity in the wire, which is sent out through power lines to run machines in our businesses and homes.

Chapter Four
Copper Compounds

Many copper compounds are ionic in nature. When dissolved in water or another solvent, these compounds come apart into ions, molecules with an electric charge.

Copper atoms are able to lose either one or two electrons from their outer electron shells. This means that copper can form more than one kind of compound. When the ion Cu^+ (a copper atom that has lost one electron) combines with another ion, the compound that it forms is called a cuprous compound. When Cu^{2+} (copper that has lost two electrons) combines with another ion, the result is a cupric compound. The differences between them can be seen by the results when copper combines with chlorine ions. A chlorine ion and a Cu^+ ion form cuprous chloride (CuCl), or copper (I) chloride, which does not dissolve in water. A Cu^{2+} ion combines with two chlorine ions to form cupric chloride ($CuCl_2$), or copper (II) chloride. Copper (II) compounds can dissolve in water to form light blue solutions, and they are much more common and stable than copper (I) compounds because the oxygen from the air can usually react with copper (I) compounds to make copper (II) compounds. The most commonly seen chemical reaction of copper is a reaction with oxygen, water, and compounds from the air to form verdigris, made of copper (II) compounds. This greenish coating protects the rest of the copper from further reactions, making copper a long-lasting roofing material. You

Chlorine ions can combine with Cu^+ ions to make cuprous chloride (CuCl). When this compound crystallizes, the molecules arrange themselves in tetrahedrons—four-sided shapes that stack like triangular blocks. Although cuprous chloride does not dissolve in water, oxygen from the air may be able to react with it. These reactions change Cu^+ ions into Cu^{2+} ions, which crystallize differently, as shown on the next page.

may have seen copper-roofed buildings and domes near your home or in photographs.

Verdigris

Bronze and brass statues and plaques are a common sight in public places in cities and towns. But not all are polished to a high gloss or protected with a thin layer of varnish or wax. The next time you see a public statue outdoors or touch a brass handrail, look to see if it is a polished

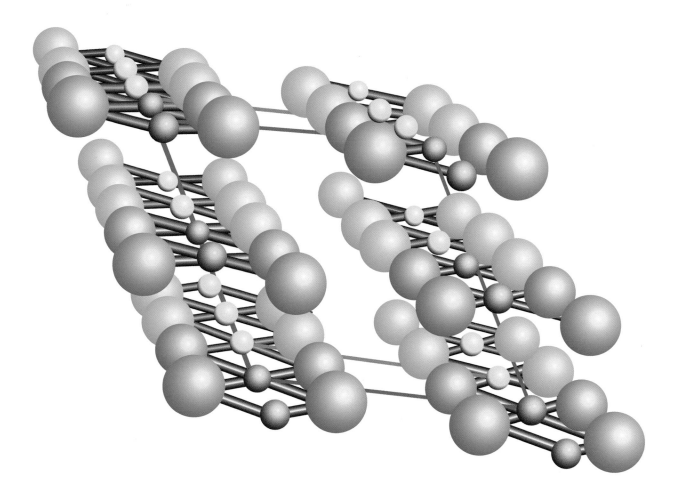

Chlorine ions can combine with Cu^{2+} ions to make cupric chloride ($CuCl_2$). When this compound crystallizes, the molecules arrange themselves in regular rows of Cu^{2+} ions between two chlorine ions. The rows are loosely attracted to each other. While cupric chloride does dissolve in water, this compound is less reactive to air than cuprous chloride. Copper (II) compounds are more common and stable than copper (I) compounds.

golden color all over, or if there is a dark patch, perhaps on the underside. That's verdigris, a compound of hydrated copper sulfate or hydrated copper carbonate. This greenish or brownish patina discolors the surface of objects made of copper and copper alloys, but it is not crumbly like rusted iron.

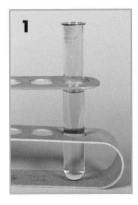

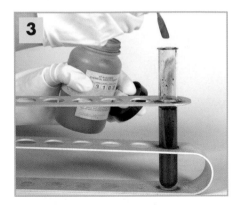

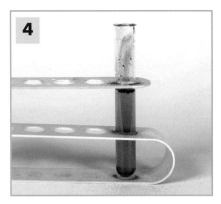

Is it hard to believe there is any copper in the blue crystals of copper sulfate? You can remove just the copper from a solution of copper sulfate. Dissolve a spoonful of copper sulfate in water in a test tube, and watch the water turn blue. Then add a spoonful of zinc or iron powder, cover the test tube, and shake. The blue color of the solution will fade and disappear, and reddish copper powder will form as a sediment on the bottom of the test tube. During this change, you may also feel the solution get warmer if you hold the test tube.

The Statue of Liberty in New York Harbor is sheathed entirely in pure copper—some 179,000 pounds (81,193 kg) of it. The statue is not polished, so its surface has taken on the classic green color of weathered copper. But it's not rusting away, as an iron statue would. The thin weathered layer keeps air and water away from the rest of the copper. That's one reason why the statue was sheathed in copper. Another reason was the symbolic meaning of copper in alchemy as a fine metal representing

The copper of the Statue of Liberty in New York Harbor has a green patina, which protects the rest of the statue from the effects of the moist air. The copper compound cupric chloride, also called copper (II) chloride, is used in fireworks to give the exploding sparks a blue color. Cupric chloride is also used for removing sulfur from petroleum during refinement. Cuprous oxide (Cu_2O), also called copper (I) oxide, does not dissolve in water but is useful for making glass and paint.

Aphrodite and femininity. The French people, who gave the Statue of Liberty as a gift to the people of the United States in 1886, personify liberty as a female symbol.

Some copper (II) compounds are soluble in water and are used as agricultural chemicals. Some are fertilizers, while others kill fungus in the soil or on plants when used carefully. Copper is also an algicide used in water purification. The use of herbicides is regulated by government agencies so that these chemicals are sold with accurate labels and instructions. Often they can only be sold by a trained, licensed clerk. Copper is an essential mineral found in vitamins and some food supplements sold at drugstores. In addition, some cosmetics contain a small amount of copper. Some brands of makeup and nail polish include copper for a shiny, metallic look, while some soaps and creams contain copper compounds that are believed to be good for skin and hair.

Chapter Five
Copper and You

Copper is an essential part of living things. It's a small part of a healthy diet, too, even though you don't see it on your plate. We don't eat it as mouthfuls of shiny pieces of metal; our bodies couldn't digest lumps of copper metal. We don't eat common copper compounds, either, because most of them would provide much more copper than we need and thus would be toxic. Instead, we usually obtain all the copper we need from food or vitamin pills, where copper ions are present in very small quantities.

Copper ions combine with other elements to make copper proteins and enzymes in plants and animals. Plants take copper atoms from the soil, dissolved as ions in groundwater, and combine the copper with other atoms from the soil and the air. As plants grow, they make complex molecules containing copper atoms inside their roots, stems, leaves, flowers, and fruit. Animals and people get all the copper they need (a very tiny amount) by eating plants. The copper we eat that is naturally present in plants can be digested in our bodies and carried in our bloodstreams to be used in cells all throughout our bodies.

Copper in Our Bodies

How much copper are we talking about? Well, for every billion atoms in a human body, only ninety-nine atoms are copper. A human who weighs

Plants absorb through their roots all the copper they need as a few ions dissolved in water. Inside plant cells, organic molecules are made and put to good use. When animals eat plants, the small amount of copper in these organic molecules is in the right compounds to be useful for the animals' bodies, too.

160 pounds, or has a mass of 75 kilograms, would have only 0.075 grams (0.0026 ounces) of copper in his or her body. All the copper in ten big people might be enough to make one tiny earring or a small, thin coin. That doesn't seem like much, compared to the other elements found in the human body, such as carbon and nitrogen. But it's the right amount for several important chemical reactions.

Copper is an important catalyst within living cells. A catalyst makes a chemical reaction happen, once all the other elements are present. It's like a trained conductor who leads a symphony orchestra of musicians. Copper catalyzes many chemical reactions, including oxidation in mito-chondria, which are small structures inside your cells that use oxygen to give your cells energy to move and work. Copper also catalyzes the for-mation of melanin pigments that color your skin, hair, and eyes.

Galvani's Studies

In the late 1770s, Luigi Galvani was teaching anatomy and physiology at the university in Bologna, Italy. One day while Galvani was dissecting the muscles of a frog hanging on a brass hook, he saw the dead frog move under his scalpel. He could make the frog move again when he touched the muscle with two different kinds of metal: a copper alloy and an iron or zinc alloy. Galvani went on to study electrophysiology. His research led to the invention of the galvanometer (which detects electric current) and the development of galvanized iron and steel. More recently, medical

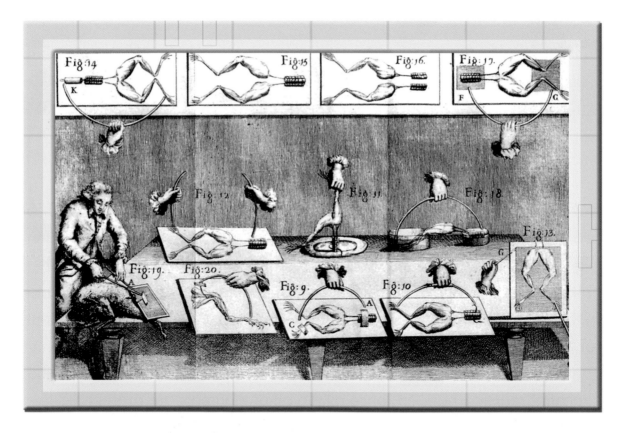

How could a dead frog move? The nerves in the frog's muscles contain fluids with dissolved ions. The scalpel that Galvani used was made of an iron alloy, but the hook was a copper alloy. The fluids connected the hook to the scalpel, causing a small electric current. This current was similar to the signals carried by the nerves to make the muscles move.

devices have been invented that stimulate the movement of muscles by weak electric shocks. You may have seen doctors and physiotherapists using these devices to help people recover from injuries.

Copper in Other Living Things

The tiny amount of copper in plants and animals does vital work to keep them alive, just like it does for us humans. There's copper in plastocyanin molecules, which are found in chloroplasts in plant cells. The chloroplasts are what make plants look green. Plants use plastocyanin and chlorophyll molecules in photosynthesis, the process whereby water, carbon dioxide, and energy from sunlight produce oxygen and sugar. Chlorophyll has a similar structure to hemoglobin, the molecule in your red blood cells that carries oxygen through your bloodstream to all the cells in your body.

Many animals, such as snails, crabs, and lobsters, don't have red blood. They use hemocyanin, a cuproprotein (copper protein) in their blood cells, to carry oxygen instead of the hemoglobin in red blood used by mammals, birds, and reptiles.

Copper Ornaments

The copper in jewelry can be very visible, in a necklace or brooch made of pure copper. You have probably seen a copper bracelet worn by an older person with arthritis. Maybe you've heard of the belief that wearing copper will reduce the pain a person feels from arthritis. While there is no scientific proof of this, some people insist they feel better when they wear a copper bracelet or take a bath in a copper bathtub.

There is no harm in this old wives' tale, because a copper bracelet doesn't cost a lot. A relaxing bath is good for people with arthritis, even without a fancy copper bathtub. But any pain relief probably comes from the placebo effect of believing in this tradition.

People who believe a copper bracelet helps their arthritis sometimes do feel better. Wearing a bracelet can be a visible reminder to help a person keep to the diet and exercise plans that help many people with arthritis. For many years before modern medicine, some illnesses were treated with powders made of gemstones or copper, silver, and gold. Alchemists believed that precious metals and stones had healing power. Some rich patients had faith in potions with expensive ingredients.

There's copper in most cheap metal jewelry. If you wear a ring that stains green around your finger, the ring has some copper in it. There's usually some nickel in inexpensive metal jewelry as well. Some people react strongly to the presence of nickel in jewelry, getting a rash or an itch from a bracelet or earring, and some react even to a little copper. There's also a little copper in most silver and gold jewelry, unless it is certified twenty-four-karat gold or 100 percent pure silver. Sterling silver is 92 percent silver, and the rest is mostly copper. Jewelry may seem unimportant because it is only decorative, but it was through making wires to craft necklaces that metalsmiths first learned that copper is malleable and ductile.

Copper in Your Tools and Home

Copper is in your body, your pets' bodies, and in plants and the soils used by plants. Copper is in food, in the tools that you use, and in your home as well. The pipes in your home that carry water to your kitchen and bathroom are probably made of copper. These pipes can be easily installed, soldered together, joined to taps (faucets) and other brass fittings, and even heated and bent around corners where necessary. The metal does not rust and taint the drinking water. Copper water pipes are not a new invention, by the way. One of the ancient Egyptian pharaohs had copper pipes installed in his bath, and those pipes were still whole and usable when they were dug up 5,000 years later.

The average modern American home contains about 400 pounds (181 kg) of copper. Some of it you can see in hinges, handles, doorknobs, and keys. Some of it is behind the walls, in water pipes as well as in electrical and telephone wiring or television cables. Anything with electrical parts has copper wiring in it, including refrigerators, computers, ovens, and microwave ovens.

You are probably wearing some copper right now. There's copper in zippers and in snaps and metal button shanks. There are copper compounds in the dyes that make clothing colorful, both natural plant dyes and modern chemical dyes.

Most forms of transportation make use of copper. The average motor vehicle has at least 50 pounds (23 kg) of copper, while a city bus has even more. For instance, there's copper wire wound into the ignition coils used to start up an engine. The new models of electric cars require more copper than gasoline-powered vehicles do. Even bicycles contain copper. And if you simply ride a skateboard, you're still using some copper—it's in the trucks and also in the shoelace eyelets on your shoes. The average North American, during his or her lifespan, will own and use products containing some 1,500 pounds (680 kg) of copper.

The Periodic Table of Elements

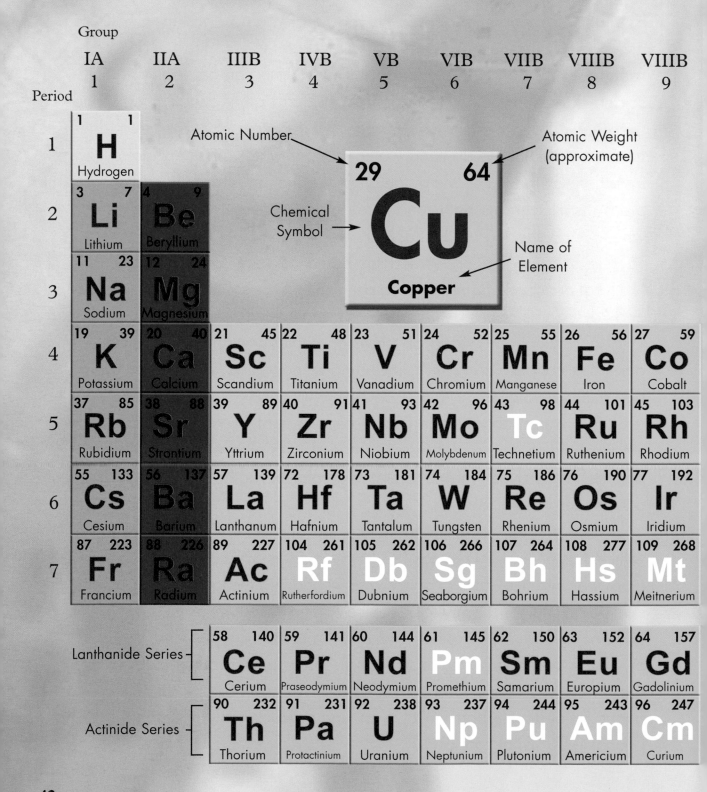

Group

| IA 1 | IIA 2 | IIIB 3 | IVB 4 | VB 5 | VIB 6 | VIIB 7 | VIIIB 8 | VIIIB 9 |

Period

Atomic Number → 29   64 ← **Atomic Weight (approximate)**

Chemical Symbol → **Cu**

Copper ← **Name of Element**

1 — 1 1 **H** Hydrogen

2 — 3 7 **Li** Lithium | 4 9 **Be** Beryllium

3 — 11 23 **Na** Sodium | 12 24 **Mg** Magnesium

4 — 19 39 **K** Potassium | 20 40 **Ca** Calcium | 21 45 **Sc** Scandium | 22 48 **Ti** Titanium | 23 51 **V** Vanadium | 24 52 **Cr** Chromium | 25 55 **Mn** Manganese | 26 56 **Fe** Iron | 27 59 **Co** Cobalt

5 — 37 85 **Rb** Rubidium | 38 88 **Sr** Strontium | 39 89 **Y** Yttrium | 40 91 **Zr** Zirconium | 41 93 **Nb** Niobium | 42 96 **Mo** Molybdenum | 43 98 **Tc** Technetium | 44 101 **Ru** Ruthenium | 45 103 **Rh** Rhodium

6 — 55 133 **Cs** Cesium | 56 137 **Ba** Barium | 57 139 **La** Lanthanum | 72 178 **Hf** Hafnium | 73 181 **Ta** Tantalum | 74 184 **W** Tungsten | 75 186 **Re** Rhenium | 76 190 **Os** Osmium | 77 192 **Ir** Iridium

7 — 87 223 **Fr** Francium | 88 226 **Ra** Radium | 89 227 **Ac** Actinium | 104 261 **Rf** Rutherfordium | 105 262 **Db** Dubnium | 106 266 **Sg** Seaborgium | 107 264 **Bh** Bohrium | 108 277 **Hs** Hassium | 109 268 **Mt** Meitnerium

Lanthanide Series — 58 140 **Ce** Cerium | 59 141 **Pr** Praseodymium | 60 144 **Nd** Neodymium | 61 145 **Pm** Promethium | 62 150 **Sm** Samarium | 63 152 **Eu** Europium | 64 157 **Gd** Gadolinium

Actinide Series — 90 232 **Th** Thorium | 91 231 **Pa** Protactinium | 92 238 **U** Uranium | 93 237 **Np** Neptunium | 94 244 **Pu** Plutonium | 95 243 **Am** Americium | 96 247 **Cm** Curium

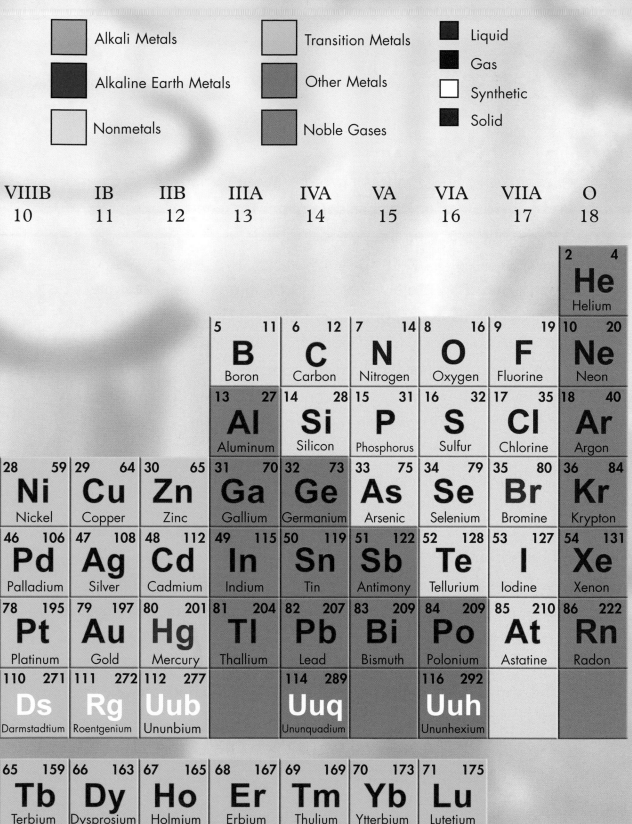

Glossary

atomic number The number of protons in the nucleus of an atom of an element. The atomic number determines an element's structure, properties, and place on the periodic table.

atomic weight Also known as atomic mass. The average of the weights (or more accurately, masses) of all the different naturally occurring forms (isotopes) of an atom of a specific element.

compound A substance that is made from the atoms of two or more elements joined together by chemical bonds.

density How much mass an object contains in a given volume (mass per unit volume), which is often represented as grams per cubic centimeter (g/cm^3).

group In the periodic table, each vertical column of elements. Elements in a group have similar properties.

ion An atom or molecule with an electric charge due to the loss or gain of electrons.

isotopes Atoms that have the same number of protons but a different number of neutrons. Isotopes are different forms of a particular element.

mass number The number of protons and neutrons in the nucleus of an atom of an element.

metalloid An element that has some properties of a nonmetal and some of a metal.

oxide A compound that includes oxygen and one other element.

oxidize To combine with oxygen.

period In the periodic table, each horizontal row of elements.

sulfide A sulfur compound that contains no oxygen.

valence electrons The outer shell of electrons that allows atoms to link together chemically and metals to conduct heat and electricity.

For More Information

Arizona Mining Association
741 East Palm Lane, Suite 100
Phoenix, AZ 85004
(602) 266-4416
Web site: http://www.azcu.org

Oregon Museum of Science and Industry
1945 Southeast Water Avenue
Portland, OR 97214-3354
(503) 797-4000
Web site: http://www.omsi.edu

Royal British Columbia Museum
675 Belleville Street
Victoria BC V8W 9W2
Canada
(250) 387-2478, 888-447-7977
Web site: http://www.royalbcmuseum.bc.ca

Web Sites

Due to the changing nature of Internet links, the Rosen Publishing Group, Inc., has developed an online list of Web sites related to the subject of this book. This site is updated regularly. Please use this link to access the list:

http://www.rosenlinks.com/uept/copp

For Further Reading

Beatty, Richard. *Copper*. New York, NY: Benchmark Books, 2001.
Gonick, Larry, and Craig Criddle. *The Cartoon Guide to Chemistry*.
　　New York, NY: HarperCollins, 2005.
Hudson, John. *The History of Chemistry*. New York, NY: Routledge, 1992.
Oxlade, Chris. *Elements and Compounds* (Chemicals in Action).
　　Chicago, IL: Heinemann Library, 2002.

Bibliography

Chemistry Plus. CD-ROM. Sunnyvale, CA: Super Tutor, 1998.
Cotton, F. Albert, and Geoffrey Wilkinson. *Advanced
　　Inorganic Chemistry*, 6th ed. New York, NY: John Wiley &
　　Sons, 1999.
Knapp, Brian. *Copper, Silver, and Gold* (Elements). Danbury, CT: Grolier
　　Educational, 1996.
Lambert, Mark. *Spotlight on Copper*. Vero Beach, FL: Rourke
　　Enterprises, 1988.
Mascetta, Joseph A. *Chemistry the Easy Way*, 4th ed. Hauppauge, NY:
　　Barron's Educational Series, 2005.
Swertka, Albert. *A Guide to the Elements*, 2nd ed. New York, NY:
　　Oxford University Press, 2002.
Tweed, Matt. *Essential Elements: Atoms, Quarks and the Periodic Table*.
　　New York, NY: Walker, 2003.

Yee, Gordon T., Jeannine Eddleton, and Chris E. Johnson. "Copper Metal from Malachite Circa 4000 BCE." *Journal of Chemical Education*, December 2004. Retrieved June 2005 (http://jchemed.chem.wisc.edu/ hs/Journal/Issues/2004/Dec/abs1777.html).

Index

About the Author

Paula Johanson has worked as a writer and teacher for twenty years. She has been nominated twice for the national Prix Aurora Award for Canadian Science Fiction. At conferences each year, she leads panel discussions on practical science (usually biochemistry) and how it applies to home life and creative work. She has written and edited curriculum materials for the Alberta Distance Learning Centre in Canada. Her raku-fired pottery and glass crafts gleam with copper accents. She lives in Victoria, British Columbia.

Photo Credits

Cover, pp. 1, 8, 12, 13, 16, 27, 32, 33, 37, 42–43 by Tahara Hasan-Anderson; p. 5 © Kaj R. Svensson/Photo Researchers, Inc.; pp. 9, 34 by Maura McConnell; p. 14 © Klaus Guldbrandsen/Photo Researchers, Inc.; p. 18 © Charles D. Winters/Photo Researchers, Inc.; p. 20 © The Museum of East Asian Art/Heritage-Images, The Image Works; p. 21 © Christos Kalohoridis/Corbis; p. 23 © Photo Cuisine/Corbis; p. 26 courtesy of NASA Jet Propulsion Laboratory; p. 28 © Jim Richardson/ Corbis; p. 29 Cat. # 345, Royal British Columbia Museum; p. 30 © Mark A. Schneider/Photo Researchers, Inc.; p. 35 © Benjamin Rondel/Corbis; p. 38 © Time Life Pictures/Mansell/Time Life Pictures; p. 40 © Mark Thomas/Photo Researchers, Inc.

Special thanks to Megan Roberts, director of science, Region 9 Schools, New York City, NY, and Jenny Ingber, high school chemistry teacher, Region 9 Schools, New York City, NY, for their assistance in executing the science experiments illustrated in this book.

Designer: Tahara Anderson; Editor: Kathy Kuhtz Campbell